Nelson Inter
Science
Student Book 4

Anthony Russell

OXFORD
UNIVERSITY PRESS

OXFORD
UNIVERSITY PRESS

Great Clarendon Street, Oxford, OX2 6DP, United Kingdom

Oxford University Press is a department of the University of Oxford.
It furthers the University's objective of excellence in research, scholarship,
and education by publishing worldwide. Oxford is a registered trade mark of
Oxford University Press in the UK and in certain other countries

Text © Anthony Russell 2012
Original illustrations © Oxford University Press 2014

British Library Cataloguing in Publication Data
Data available

978-1-4085-1723-9

11

Printed and bound by CPI Group (UK) Ltd, Croydon, CR0 4YY

Acknowledgements

Cover illustration: Andy Peters
Illustrations: Maurizio de Angelis, Tony Forbes, Simon Rumble and Wearset Ltd
Page make-up: Wearset Ltd, Boldon, Tyne and Wear

The authors and the publisher would like to thank Judith Amery for her contribution to the
development of this book.

The authors and the publisher would like to thank the following for permission to reproduce
material:

p.13: Kumar Sriskandan/Alamy; p.15: (top left) HolidayVisionStudio/Fotolia, (top middle)
hotshotsworldwide/Fotolia, (top right) Synelnychenko Dmytro/Fotolia, (middle) kawisphoto/
iStockphoto, (bottom left) David Hosking/FLPA, (bottom right) nutsiam/Fotolia; p.19: (ant)
andrey Pavlov/iStockphoto, (bird) Worakit Sirijinda/iStockphoto, (dog) Eric Isselée/Fotolia,
(earthworm) john shepherd/iStockphoto, (fish) The Dragon/Fotolia, (frog) TessarTheTegu/
iStockphoto, (housefly) arlindo71/iStockphoto, (human) Glenda Powers/Fotolia, (snail)
Ursula Alter/iStockphoto; p.21: (crab) Kaiya_Rose/Fotolia, (spider) Colette/Fotolia; p.22:
Alexey Klementiev/Fotolia; p.24: (left) Steve Trewhella, (right) Photo Researchers; p.25: (top)
Press Association Images/AP, (bottom) Worakit Sirijinda/iStockphoto; p.27: Johan Ramberg/
iStockphoto; p.30: (top) Beboy/Fotolia, (middle) Melking/Fotolia, (bottom) Mulden/Fotolia;
p.35: (left) koya79/Fotolia, (middle) NilsZ/Fotolia, (right) L-amica/Fotolia; p.38: Cristian Baitg/
iStockphoto; p.46: Elnur/Shutterstock; p.54: Josh Friedman/iStockphoto; p.67: -M-I-S-H-A-/
iStockphoto; p.74: (left) Pat on stock/Fotolia, (top right) Marc Dietrich/Fotolia, (bottom right)
JcJg Photography/Fotolia; p.75: (left) Maris Zemgalietis/iStockphoto, (middle) David Thorpe/
Alamy, (right) kocetoilief/Fotolia; p.84: (bottom left) Frédéric Prochasson/Fotolia, (bottom
right) rabbit75_fot/Fotolia, (top) Jakub Sliwa/Corbis Images.

Contents

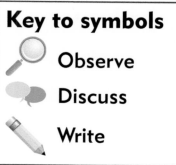

Key to symbols

Observe

Discuss

Write

Skeletons

Humans, and some animals, have bony **skeletons** inside their bodies.

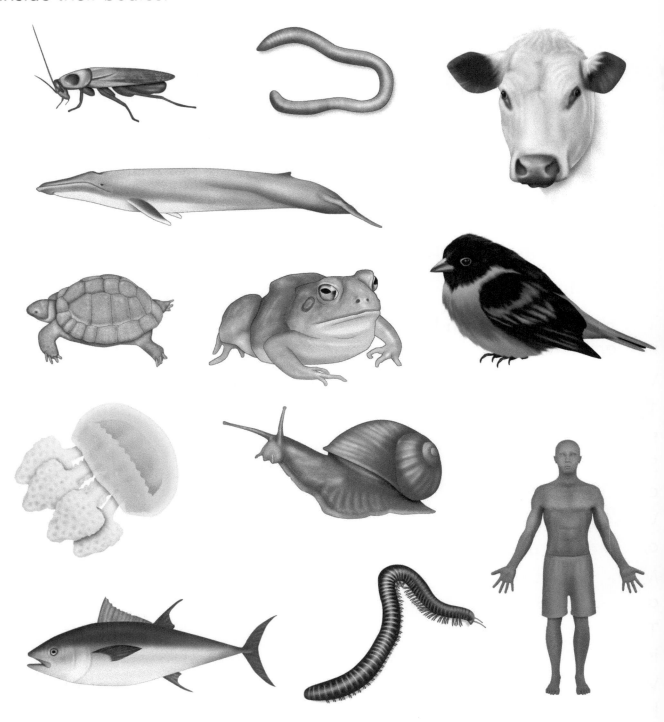

Activity 1

You will need: paper (or Workbook) and a pen or pencil.

 1 Look at the animal pictures. You can make two groups. For example, the cow belongs in group A and the snail belongs in group B.

 2 Try to sort out the rest of the animals into these two groups. Which animals belong with the cow (group A), and which with the snail (group B)? Write them down.

 3 Show your groups to the people you are working with. Discuss how you have sorted the animals.

4 Share the groups with the class.

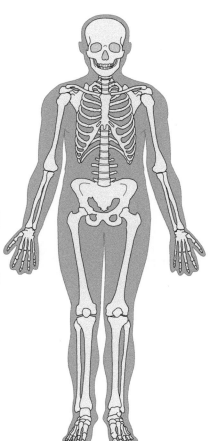

The animals in one group have a bony skeleton inside their bodies – an **internal** skeleton.

Which group is that: A or B?

The animals in the other group have no internal skeleton.

Some animals in this other group have a skeleton outside their bodies – an **external** skeleton.

Which three animals in the pictures have this kind of skeleton?

Many animals have no skeleton of any kind. Which animals in the pictures have no skeleton?

The human skeleton is the frame for our body. It is inside our **muscles**.

We can feel parts of our skeleton in our arms, legs, hands, feet and head, as well as other places.

The human skeleton is made of 206 bones.

Functions of the skeleton

The skeletal system is made up of bones, muscles and **joints**. As our bodies grow, our bones and muscles also grow. Our skeleton must grow bigger and stronger to **support** our heavier and larger body. The bigger the body, the bigger the frame supporting it needs to be.

The skeleton also allows us to move. (We will look at this in more detail below.)

In addition to allowing movement and support, it also helps to keep parts of our body safe. It **protects** certain vital organs and it is essential for breathing.

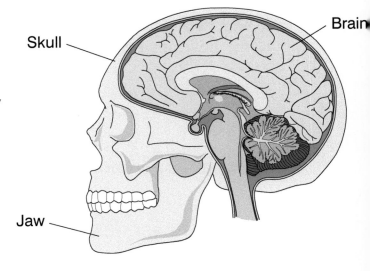

Skull

Brain

Jaw

The **skull** is a strong hard box in which the delicate brain is safely housed.

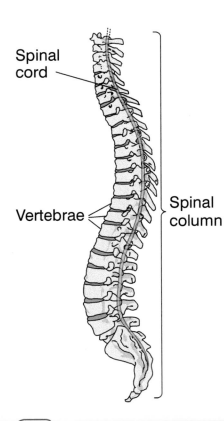

Spinal cord

Vertebrae

Spinal column

The spinal cord, which is a continuation of the brain, is also enclosed in a column of small bones called the vertebrae. Together these make up the wonderfully flexible **spinal column** (backbone or spine).

The **ribs** do two jobs:

- they allow the chest to be expanded and contracted so that air can be drawn into and squeezed out of the lungs.
- they also help to protect the lungs and the heart. The heart is located under the breastbone, where the ribs come together at the front of the chest.

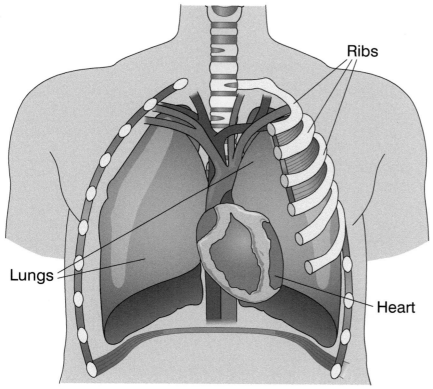

Ribs

Lungs

Heart

Inside some bones there is red **marrow**, which makes new blood cells. There are three kinds of blood cells: red blood cells, white blood cells and platelets.

It is essential that the marrow goes on producing new cells throughout our lives, because blood cells do not live long and must be replaced constantly.

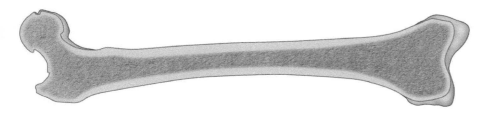

Muscles and bones

Muscles without bones would not be able to support our body, or move it from place to place.

Bones are hard and strong, and where two bones meet in a joint, there is a slippery surface on the end of each bone. This allows the bones to move smoothly over one another in the joint. The bones in the joint are held together by **ligaments**, which are made of strong elastic tissue.

Bones and joints cannot produce movement on their own. It is the muscles that move the joints and this leads to body movements and **locomotion**.

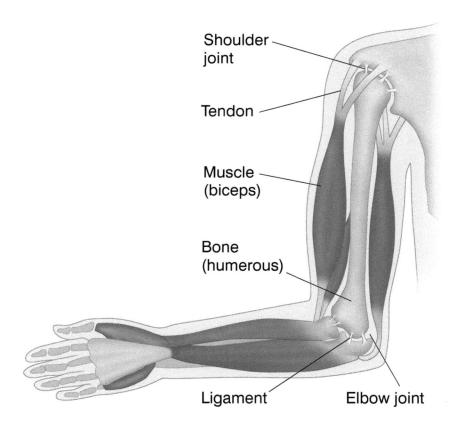

Shoulder joint

Tendon

Muscle (biceps)

Bone (humerous)

Ligament   Elbow joint

The muscles are attached to the bones by **tendons**.

Movement of bones

The skeleton, and the muscles attached to it, are essential for movement and for locomotion. A rigid skeleton without joints would not allow any movement.

When a muscle **contracts**, it pulls on the tendon, which pulls on the bone. The result is some movement.

Muscles act in pairs, pulling on the bones from one side or the other. This is what happens every time we walk, pick up a pencil, kick a ball, swim or eat. The skeletal and muscular systems – the bones, joints and muscles – have to work together to produce movement.

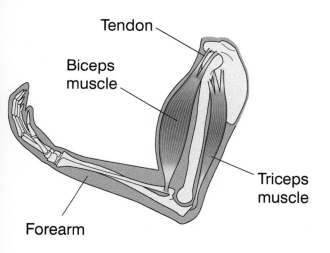

Tendon
Biceps muscle
Triceps muscle
Forearm

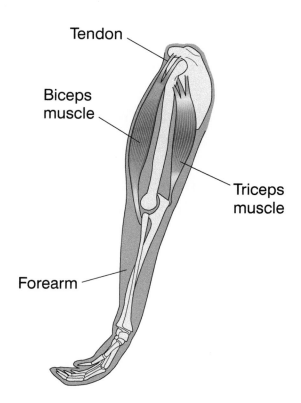

Tendon
Biceps muscle
Triceps muscle
Forearm

Drugs as medicines

A **drug** is any substance, other than food, that causes changes in the body. Drugs can be swallowed, breathed in, injected or applied to the body in some other way. There are three groups of drugs:

- **Prescription** drugs
- Over-the-counter drugs
- **Prohibited** drugs.

Activity 2

You will need: paper (or Workbook) and a pen or pencil.

1 Discuss the three groups of drug types. Share your ideas about why some drugs are in one group and not in another.

2 Write down examples of each group of drugs.

3 Share your lists and ideas with the class. Add to your lists any extra examples given by your classmates.

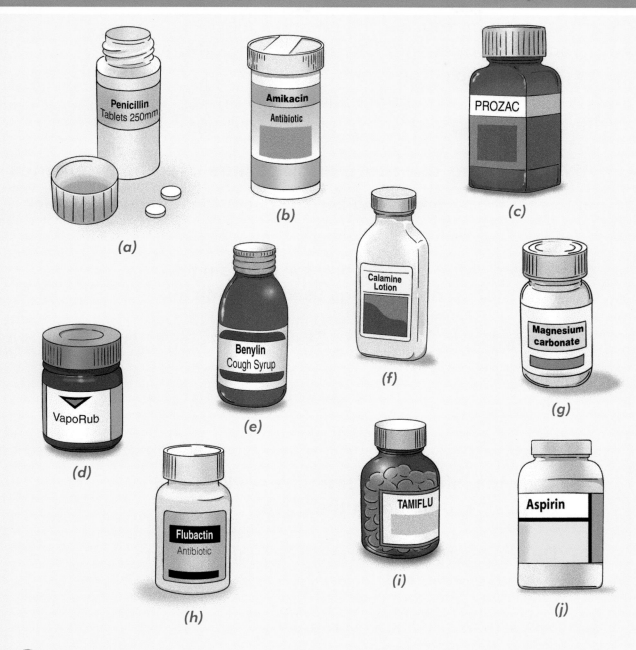

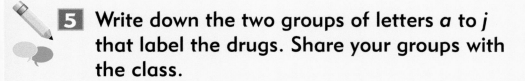

4 Look at the pictures of various drugs and sort them into two groups:
  a Prescription drugs
  b Over-the-counter drugs.

5 Write down the two groups of letters *a* to *j* that label the drugs. Share your groups with the class.

Drugs bought in **pharmacies** and chemists all have very important information on their packaging. This is given to protect people from harm. Drugs are dangerous. This is true of prescription drugs and over-the-counter drugs, as well as prohibited drugs.

Activity 3: Finding out more about drugs used as medicines

You will need: drug packets and bottles, paper (or Workbook), a ruler and a pen or pencil.

1 Draw a **table** like this one for recording information about drugs used as **medicines**.

Drug	Use	Dosage	Expiry date	Side effects	Warnings
A					
B					
C					

2 Copy the information from the packets and bottles, recording it in the table.

3 Decide how you will present the information you have collected.

4 Sort the drugs out in some way. For example, create:

   a groups for each type of drug use

   b groups for each type of warning.

5 Display the drug containers.

6 Share your findings with the class.

    a Discuss what you have all found out about the proper use of drugs as medicines.

    b Design a poster or web page about the safe use of drugs as medicines.

To use any drug safely, we must follow the instructions about the dose, the age of the patient, and the warnings about any side effects.

Taking more of a drug does not mean we will gain more benefit from it. It might lead to serious damage to our bodies, or even death. For example, organs such as the liver and the kidneys can be damaged by high drug doses and, if the damage is very bad, it might not be possible for them to recover.

Children are in the greatest danger from drugs, because their bodies are smaller and so they are more easily damaged by an overdose of drugs. This is why drugs must be stored safely at home, in a place where young children cannot reach them.

Activity 4

You will need: paper (or Workbook) and a pen or pencil.

1 Discuss with your group:

a 'What are the benefits of drugs?'

b 'What are the harmful effects of drugs?'

2 Keep notes of your group's answers to these two questions.

3 Share your group's answers with the class.

Prescription and over-the-counter drugs have many benefits.

Some can cure diseases, killing the organisms that have invaded our bodies.

Others can take away symptoms, such as the sneezing and coughing of a flu infection, without killing the virus causing the flu.

Some can prevent us being infected. They can give us protection against disease. Anti-malarial drugs are examples of this kind; they can protect us from getting malaria.

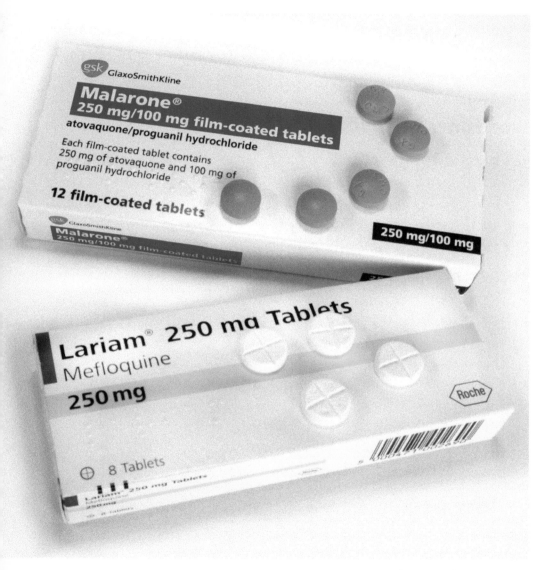

Anti-malarial drugs

If our body is not working properly it can cause a disease such as diabetes. Drugs can sometimes correct such a fault and allow the person to go on living as before.

Animals in their habitats

Your school is surrounded by **habitats**. Some are very small and others are enormous. A building itself creates habitats for certain animals. Trees, especially when they are old and large, also provide habitats for some animals and even other plants. Each **environment** has many, many different habitats for its animals and plants.

Activity 1: Investigate two different habitats

You will need: two different habitats, paper (or Workbook) and a pen or pencil.

1 Discuss with your group which **two** habitats you will **investigate** in your locality. Try to choose two that are very different.

2 Plan what you will look for and how you will **record** what you **observe**.

> **Remember** that drawing, measuring, counting and writing notes are all part of data collection and recording.

3 Go outside with your notebook and pen and investigate your chosen habitats.

a Collect enough information to be able to give a clear and full description of the animals and their habitats to the class.

b Take care to describe what each habitat is like. For example:
- Is it hot and dry?
- Is it shady?
- Is it wet?
- Is it bare soil, rock or sand?
- Is it in or near water?
- Are there plants, and are they close together or spread out?

4 **Return to the class.**

  a Discuss with your group the information you have collected.

  b Decide how you will present your descriptions to the class.

5 **Present your descriptions and then answer questions from others in the class.**

Look at the pictures.

Choose one of the animals.

Discuss what the habitat of your chosen animal is like.

What is special about the place it lives in?

Describe it to the class.

There are many different natural habitats on Earth.

Some are so cold that very few things can live there: for example, the Arctic and the Antarctic.

Others are so hot and dry that very few plants or animals can live there: for example, deserts in South America, Africa and Asia.

Very few people live in these extremely cold and extremely hot places. It is too difficult to stay alive.

Plants and animals can be found in most places on Earth. Each place has its special animals and plants that 'fit' the conditions. Usually the amounts of water, light and heat found in each place are the most important features of any habitat.

Look at the animals in the pictures and discuss what is wrong with the habitats they are in.

Tell the class what you think.

Each of the animals in the pictures above is shown in its natural habitat.

Use these words to complete the sentences on page 18:

> tadpoles   tiger   hide   feed   warm   cold
> frogs   slide   smooth   water   flowers
> snakes   eggs   birds   penguins   sea   shape
> fish   nest   bees   pond   deserts

Copy and complete these sentences using the words from page 17.

1 The _____ lives in the jungles of India. Its stripes help it to _____ when it is hunting.

2 _____ live in large numbers in the _____. They swim and _____ together. Their _____ helps them to move easily through the _____.

3 The Antarctic is a very _____ habitat so the _____ and other animals have to have a way of keeping _____. The feathers protect the _____ from the cold.

4 Some _____ live in very hot, dry places such as _____. Their skin is dry and it is very _____ which helps them to _____ over the sand and rocks.

5 _____ and other insects visit _____ to collect food. They must use it for themselves or take it back to their _____ to feed their young.

6 A _____ is a good place for _____ because they must have fresh water to lay their _____ in. Adults can come out of the water, but the eggs and _____ must live in it as they grow.

Ant

Earthworm

Frog

Human

Fish

Housefly

Snail

Dog

Bird

Activity 2: Sorting and identifying animals by their features

You will need: paper (or Workbook) and a pen or pencil.

1 Discuss with your group what you remember from Stage 3 about grouping animals, using their simple features – for example, body covering or colour.

Continue over the page

2 Look at the animals shown in the pictures on page 19 and discuss how they could be sorted.

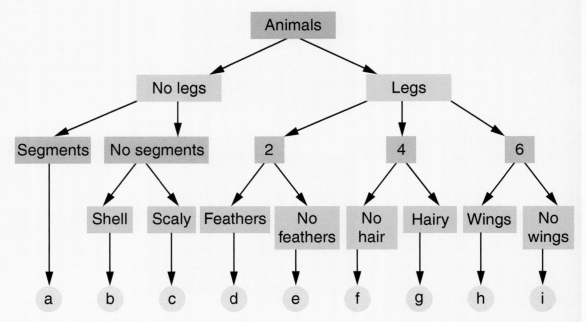

An identification key

3 You can use the **key** above to identify the animals in the picture.

> **Remember** to use the simple features you can see to help with the identification.

 4 Write down the letters (a) to (i) and when you have identified the animals, write their names beside the correct letters.

 5 Share your **results** with the class.

Crab

Spider

6 If you had to include a spider and a crab in the key, what would you add to it?

a Discuss your ideas with the group.

b When you have an answer, share it with the class.

7 Choose another animal and ask your group to fit it into the key. Share your group's ideas with the class.

Keys are a very useful way to sort and identify living things – plants as well as animals. When identifying a plant, it is useful to be able to look at flower shapes and colours, leaf shapes and sizes, types of stem and fruits. Simple features of animals, such as the number of legs, body covering, or method of reproduction are useful when identifying them.

Human activity and the environment

The environment is changed by human activity. These changes are sometimes good, but they can also be bad, doing damage to the living and non-living parts of the environment.

This litter will not rot and most of it will end up in landfill sites

Activity 3

You will need: an area outside, paper (or Workbook) and a pen or pencil.

1. Go outside with your group and choose a place where you can do a survey of litter on the ground. It might be inside or outside the school playground.

2. Keep a tally of each kind of rubbish you find (for example, plastic bags or drinks cans).

3. Use the data to make a bar graph. Display your group's graph with those from other groups.

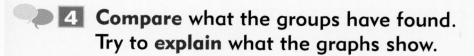

4 **Compare** what the groups have found.
Try to **explain** what the graphs show.

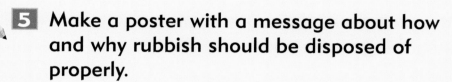

5 Make a poster with a message about how
and why rubbish should be disposed of
properly.

Display your posters around the school.

People produce litter and other wastes. If the
environment is dirty and littered, then people are to
blame. Animals and plants don't create litter.

The answer to the problem of waste **disposal** is for us
human beings to work out.

Activity 4: What are the effects of litter?

You will need: paper (or Workbook) and a pen or pencil.

1 Look at the pictures and discuss what effects such things have on:

   a people

   b other living things

   c non-living parts of the environment.

2 Write lists of all the effects that the group can think of.

3 Display your lists under the headings:

   people   other living things   non-living things

Careless disposal of waste products causes damage. It can also cause the spread of diseases among people, through **pollution** of our water supplies, and through the flies and rats that feed on some types of waste.

Other living things are also damaged, especially if water becomes poisoned with wastes from factories. Such wastes in rivers and in the sea kill plants and animals living in the water. If oil is dumped or spilled on the sea, many birds and mammals can become covered in oil and this kills them.

Smoke from factories should be cleaned before it is released into the air. If it is not, it can damage people's lungs and it can pollute the rain. When this dirty rain (also called **acid rain**) falls, it kills trees and other plants. It also poisons lakes and rivers, killing the fish and other animals.

Beautiful places are spoiled by people's rubbish. Beaches should be clean and safe, but when people on land and on ships throw their rubbish into the sea, a lot of it ends up on the beaches. Some of it is dangerous; all of it is ugly.

Sewage disposal is another big problem, especially when towns and cities grow bigger and bigger.

It is a mistake just to pump sewage into the sea. Sewage carries diseases. It also encourages flies and rats to breed, and they spread diseases. If the ground water becomes polluted with sewage, all the people taking water from the well or the pump can become ill.

The same is true of rivers and streams. It costs money to build and operate sewage treatment works, but it is very important.

Unfortunately, in real life, not everyone takes care of the environment. Individuals, families, communities and even whole countries are sometimes selfish and careless. They do damage by the way they behave. Sometimes it does not harm them, but it can harm other people and places far away.

Sometimes the damage happens slowly over a long period of time. For example, air pollution from burning oil, petrol, coal or wood.

This picture shows one way to help protect the environment.

What do the children use the containers for?

What could they do with the different materials they collect?

Activity 5

You will need: bins, paper (or Workbook) and a pen or pencil.

1 Discuss with your group what materials you will **recycle** at school.

2 Make bins for each of the chosen materials. Put a label on each one.

3 Write instructions to tell the class how to use the bins. Display the instructions near the bins.

People must care for the environment. If we do not care for it, it will be damaged. Plants, animals, soil, water, even the sea, can all be spoiled.

One big danger is from the rubbish we produce. If we just throw it away it spoils the environment. It is also very wasteful. Many materials can be recycled or used again.

Some materials – from plants and animals – can be put back into the soil. They will **decay** and produce **minerals** that plants can use for growth.

Paper, which is made from wood, can be recycled to make cardboard and other low-quality papers.

Metals and glass can be melted and recycled to make new items.

Some plastics can also be recycled, but some cannot. If they are just thrown away they will stay in the sea or on the land for a very, very long time.

If possible, find out about the local arrangements for **recycling** wastes so that your class can really become active in protecting the environment through recycling.

The Earth is a beautiful place and we must all share the responsibility of looking after it.

Chapter 3: States of matter

Solids, liquids and gases

ice

clouds

rain

snowy peaks

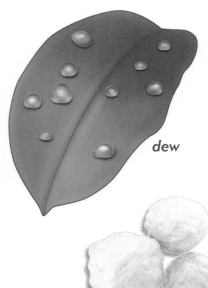

dew

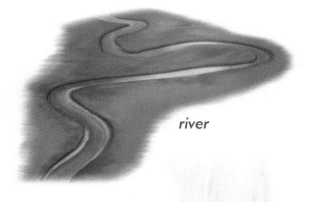

river

hailstones

boiling water

Activity 1: Identifying water in its three states

You will need: paper (or Workbook) and a pen or pencil.

1 **Look at the pictures on page 28.**

   a Choose the ones that show water.

   b Tell the class which ones you have chosen.

2 **Draw a table like this.**

Solid	Liquid	Gas

   a Identify the form of water in each picture.

   b Record your answers in the columns under the correct headings.

3 **Copy and complete the sentences below.**

   Here are the words you will need:

   > **liquid   states   gas   water   solid**

   (You will need to use some words more than once.)

   Water is found in three _____.
   They are called _____, _____ and _____.
   The _____ state is called ice.
   _____ is the _____ state.
   The _____ state is called **water vapour**.

All **matter** in the universe exists in these three **states**: **solid**, **liquid** or **gas**.

Water is the most common example on Earth of a substance that can be easily found in all three states.

The rocks and soil of the Earth's crust are solid, but deep inside the Earth they are in a liquid state. Why?

The oceans covering much of the Earth are mostly liquid water, with some solid water (ice) floating on them – for example, in the Arctic Ocean. Why?

The atmosphere surrounding the Earth is mostly a mixture of gases, with some clouds of liquid water droplets floating in it. Why?

In a group discuss your ideas about the three questions above.

Share the group's ideas with the class.

Look at the four pictures and describe what is happening to the water in each one. Try to explain what you have described. Share your ideas with the class.

Look at the two pictures.

Why does your skin dry quickly after swimming on a hot day?

Which bucket 'loses' water more quickly and why?

In a group discuss your ideas and then share them with the class.

Activity 2: Experiment to test the speed of drying pieces of cloth

You will need: water, pieces of cloth, scissors, paper (or Workbook) and a pen or pencil.

1 Investigate the drying of two pieces of cloth.

   a Plan how you will investigate the drying of two pieces of wet cloth.

   b Make it a **fair test** so that you will be able to compare the results.

2 **How will you treat the two pieces of cloth?**

a Discuss with your group how you will treat the two pieces so that one dries faster than the other. Think about washing put out to dry.

b **Predict** which piece will dry faster and write down your prediction.

3 **Decide how you will measure the dryness of the cloths.**

Keep a record of what you do, especially the time when you start and finish the investigation.

4 **Discuss and compare the results.**

a Record what has happened to the cloths.

b Discuss the results with your group.

c Compare them and come to a conclusion based on your **evidence**.

d Compare the results with your prediction.

5 **Share your results with the class.**

We can change liquid water into the gas state (water vapour) by heating it. We do this when we hang washing out in the sun to dry. It happens to our wet skin after washing or swimming. It happened to the pieces of cloth in our investigation of drying.

The heat from the sun is like the heat from a fire. Our bodies make heat too.

When wet things – for example, our skin, dishes, fabrics or puddles – become dry, the liquid water has been changed into water vapour.

This process of changing from liquid to gas is called **evaporation**. The gas is **invisible**, so the liquid just 'disappears' into the air.

(a)

(b)

(c)

Look at the pictures. Each one shows what happens when water vapour (an invisible gas) is cooled down.

a The water vapour in the warm air touches the cold window and the water vapour is cooled and changes back into liquid water. It forms tiny drops of water on the glass as it **condenses**. This is called condensation.

b The water vapour in the warm air touches the cold can and the water vapour is cooled and changes back into liquid water as it condenses.

c Try to explain the clouds in this picture. Tell the class what you think.

Some other common materials can also have their state changed easily.

Activity 3: Investigating a material's changes of state

You will need: a heat source; wax, chocolate, butter or margarine; paper (or Workbook) and a pen or pencil.

1 Choose which material you will investigate: wax, chocolate, butter or margarine.

⚠ **WARNING:** Take care when using the heat source to change the state of your material.

2 How will you change the state of your chosen solid material?

a Discuss how you will do it with your group.

b Write down what you plan to do.

3 Can you change the material back to its original state? How?

a Discuss how you will **reverse** it with your group.

b Write down what you plan to do.

Continue over the page

4 **Keep a record of what you do.**

   a Draw a picture of your chosen material before you try to change its state.

   b Record your observations of changes as they happen.

5 **When the substance has changed from the solid state you started with, keep a record of what happens.**

   a Make a second drawing of its new appearance and note any other differences.

   b Now try to reverse the change so that you return the material to its solid state.

   c Record observations of changes as they happen.

6 **Discuss with your group what you conclude about why materials change their states. Write down the group's conclusion.**

7 **Share your drawings and conclusion with the class.**

Heating solids can change their state to liquid.

When the wax, butter, chocolate or margarine were heated, they became liquids.

This is a second example of heating causing a **change of state**. The first example, when liquid water was heated, causing it to evaporate, changed it into an invisible gas (water vapour).

When the materials in Activity 2 were in the liquid state, the change from solid to liquid was reversed by letting them cool. The liquid wax, chocolate, butter and margarine all turned solid once you stopped heating them. Cooling reversed their change of state.

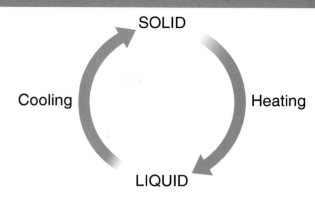

SOLID

Cooling   Heating

LIQUID

Look at the pictures and think about your investigation and the conclusion you reached.

What are the pictures telling you about changing the state of water – from liquid to solid and from liquid to gas?

Tell the class what you think.

Melting and freezing

Melting is the process of changing from the solid state to the liquid state. The materials you investigated – wax, chocolate, butter or margarine – all melted when you heated them.

In the picture on page 37, there is an example of the reverse process – changing a liquid into a solid by **freezing** it.

Cooling a liquid slowly at room **temperature** can also change its state, as you saw in the investigation. For example, liquid wax cools to become solid wax.

Freezing is just a faster and greater cooling method, where the temperature of the material is lowered far below room temperature. We do this in freezers as a way of making ice and storing foods.

Activity 4: Investigate what happens to water when it boils

You will need: paper (or Workbook) and a pen or pencil.

Your teacher will use a kettle, a plate and a bowl to demonstrate another process.

⚠️ **WARNING:** When the water boils take care not to get too close, or to put your hand near.

1 Look carefully at the spout and see if you can observe the 'gap' between the cloud and the spout.

2 Now look at the cloud.

a What do you think it is – gas or liquid?

b Tell the class what you think.

3 Watch carefully as the teacher holds the cold plate with the cloth and puts it into the invisible steam.

The bowl is standing below the point where the plate is being held.

Continue over the page

4 a What do you observe on the plate?

b What do you observe in the bowl?

c Try to explain your observations to the class.

The **boiling** water turned into **steam**, which was an invisible gas. The 'gap' was the invisible steam – the product of boiling water.

When the steam came out of the spout, it was cooled down and condensed into a cloud of tiny water droplets. The water droplets were **visible**.

Now copy and complete the sentences below.
Here are the words you will need:

> air see water clouds vapour ran
> condenses drops dripped cools liquid

(You will need to use some words more than once.)

I saw _____ of _____ on the plate. It _____ down the plate and _____ off into the bowl.

Water _____ in the air _____ down as the _____ goes higher up.

As it cools, it _____ into tiny _____ of _____ water.

We can _____ the drops of _____ as _____ in the sky.

Evaporation – the change from a liquid to a gas – can happen either slowly or quickly. The speed of change depends on the rate at which the liquid is heated.

- Boiling water is evaporating quickly. It is being heated to its boiling point and at that point it changes to the invisible gas called steam. The steam leaves the surface of the boiling water and rises into the air, where it cools, condenses and forms a visible cloud of water droplets.

So what we see above the boiling water is *not* steam. It is a cloud of water droplets, just like the clouds in the sky (and they are definitely *not* steam!).

- Most water evaporates slowly, turning into another invisible gas called water vapour. This happens at much lower temperatures than the boiling point. Water does not have to be boiling for it to evaporate.

Sound making and measuring

Listening to sounds helps us know what is going on around us. But where do sounds come from?

Activity 1: Where do sounds come from?

You will need: an area outside, paper (or Workbook) and a pen or pencil.

1 Discuss with your group where sounds come from.

2 Go outside.

   a Listen for sounds.

   b Name as many sources of sounds as you can.

3 Write down each of the sounds you hear.

4 Share them with the class.

(a)

(b)

(c)

(d)

(e)

(f)

(g)

(h)

(i)

(j)

(k)

(l)

(m)

Activity 2

You will need: paper (or Workbook) and a pen or pencil.

 1 **Look at the pictures on page 43.**

    a Discuss them with some of your classmates.

    b Are they all sources of sound?

2 **Sort them into groups:**

    a natural sources of sound

    b artificial sources of sound

    c non-sound sources.

 3 **Use the letters on the pictures to record your groups.**

4 **Share your answers with the class.**

The world is a very noisy place! We are surrounded by many sources of sound. Even our own bodies make sounds, such as the natural sounds of: **speaking**, **crying**, **laughing** or **sneezing**. Can you think of any more?

Many other sounds are natural.

Some sources of natural sound are shown in the pictures on page 43. Can you think of any more? Share your sources with the class.

Activity 3: How are sounds produced?

You will need: a wooden or plastic ruler, paper (or Workbook) and a pen or pencil.

1 Take a ruler or flat strip of wood or plastic.

   a Place it on the edge of the desk so that more than half of it sticks out over the edge.

   b Hold it down firmly.

2 Flick the free end of the ruler. Repeat this several times and listen carefully.

3 Observe what happens to the ruler when you flick it – use your senses of sight, hearing and touch.

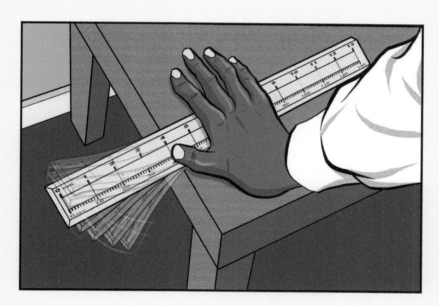

4 Record your observations.

5 Discuss your results with the group.

   a Come to a conclusion about how the sounds were produced.

   b Share your conclusion with the class.

Your observations were of three kinds:

- what you could *see* happening
- what you could *hear* happening
- what you could *feel* happening.

The ruler moved. You could see it going up and down when you let go of it.

It **vibrated**.

As it vibrated, you heard a sound. When the **vibrations** stopped, the sound stopped.

As it vibrated you could feel the movement through your fingers.

All sounds are the result of vibration. Sometimes you can see the vibrating object or material. Stringed instruments are a good example, and so are drums or thumb pianos.

Activity 4

You will need: an object you can blow into or across to make a sound, paper (or Workbook) and a pen or pencil.

1 Choose an object that you can blow into – a **wind instrument**, tube or pipe – or blow across the open top of a glass or plastic bottle.

2 Try to make sounds with your chosen object.

a Ask your group to observe the sound-making – look, listen and touch.

b Write down a record of what the group observes.

Continue over the page

3 Observe the sound-making of other groups. Record what you observe.

4 Compare the records made by the group members.

a Try to reach a conclusion based on the evidence of your observations.

b Be ready to tell the class what you heard, what you saw and what you felt.

Many vibrations are not seen, but the sound is heard. Wind instruments, such as the flute, can be heard, but the vibrations cannot be seen.

It is the air inside the instrument that is vibrated when the player blows into it. We cannot see the vibrating air, but we can hear the sound it produces.

If we touch the sound maker when it is being blown, we can sometimes feel it vibrating. This is caused by the air vibrating inside it as the player blows into it.

Activity 5: Measuring sounds using a sound-level meter

You will need: a sound-level meter, paper (or Workbook) and a pen or pencil.

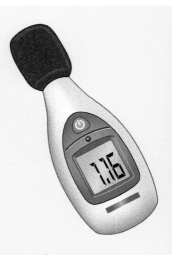

1 Handle the **sound-level meter** carefully.

a Explore how it is switched on and off and how it is used to measure the **volume** of a sound.

b Look at how you can see the measurement readings. They are in **decibels** (written **dB**).

2 When you are familiar with the meter, move around the room and outside.

a Take measurements of sounds.

b Record your observations in the correct units.

3 Plan how your group will test making a sound louder and louder, using the meter to measure the level each time. It must be a fair test. When you have a plan, show it to your teacher.

4 Do your **test** and record the readings from the sound-level meter each time you measure the sounds you make.

5 When you have finished the test, look at your data.

a Discuss them with your group and come to a conclusion about the sounds you made.

b If you can use the data to produce a **bar chart** or a table, then do it, ready to share with the class.

A table could be used, like this example.

Sound source e.g. dropping a book on the desk	Sound level (dB)
From 5 cm	x
From 10 cm	y

Or a bar chart could be used like this example.

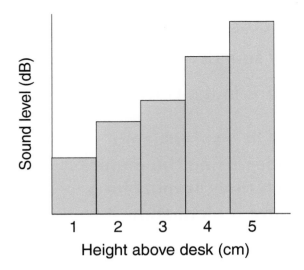

The sound level meter should have shown you that the more **energy** was put into making a sound, the higher the reading on the display.

You may have done this by hitting something harder and harder, or blowing into it harder and harder, while keeping the sound-level meter at the same distance from the source of the sound.

If you moved the sound-level meter closer and closer to the sound source, without changing the volume, then it should have shown you that the closer it was to the source, the louder the sound being measured.

The reading will have gone higher and higher as the sound-level meter was moved closer and closer to the source.

Sound travelling

Sound is one form of **kinetic energy** – energy causing movement.

Vibrations travel away from the source as sound waves. This is called the **transmission** of sound.

If sound waves reach our ears, they will set our ear drums vibrating and that is how we hear the sound.

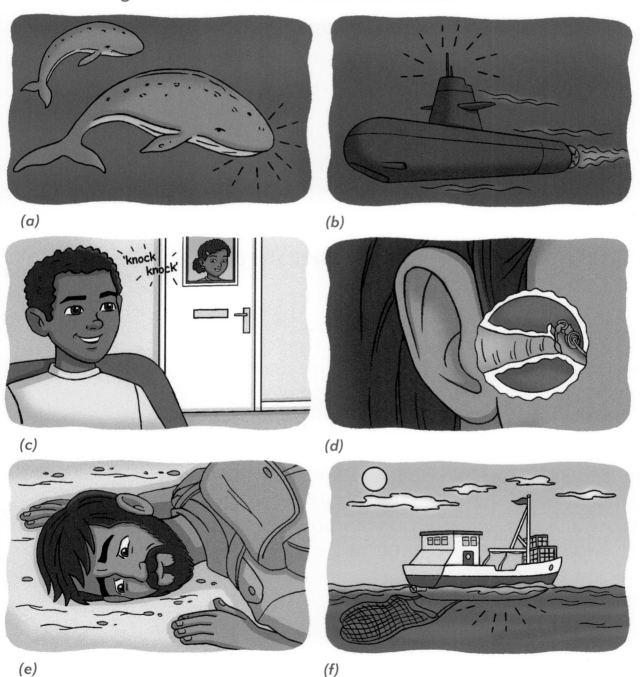

(a)

(b)

(c)

(d)

(e)

(f)

Sound waves can travel through the air. If they did not, we would not hear when our friends spoke to us, or when the radio was switched on.

Activity 6: Understanding more about the transmission of sound

You will need: paper (or Workbook) and a pen or pencil.

 1 **Look at the pictures on page 51.**

     a Discuss with your group what they show.

     b Try to work out what they have to do with the transmission of sound.

2 **For each situation, decide if sound is being transmitted and, if it is, what is it travelling through – a solid, a liquid or a gas?**

 3 **Share the group's ideas with the class.**

Sound can travel through solids.

We all know that when people knock on a door, the people on the other side can hear the sound of the knocking. If we live or work in a building with more than one floor, we can hear people walking or moving things around in the rooms above us.

Traditional hunters all over the world worked out long ago that the sound of animals moving could be heard through the ground. They understood that putting an ear close to the ground would allow them to hear animals they could not see. This was, and still is, very useful when hunting.

Activity 7: What happens to sounds when they travel through a solid?

You will need: a desk or table, a stick, paper (or Workbook) and a pen or pencil.

1 Work with a partner for this activity.

a Sit at opposite ends of the desk or table.

b Cover one ear with a hand and put your other ear close to the desk top, but *not* touching it.

c Ask your partner to lightly scratch the desk top with a pencil or stick.

d Now put your ear closer to the desk so that it *is* touching the desk top.

e Ask your partner to repeat the scratching.

Continue over the page

2 Swap with your partner and repeat the activity.

3 Discuss what you both heard and come to a conclusion.

Sound can also travel through liquids, such as water.

Whales that swim over long distances in the deep oceans need to be able to communicate with one another. This is especially true for mothers and their young. Whales send out sounds that travel through the sea and can be heard over long distances. They also catch their food by using sound to find the fish or other sea creatures that they feed on.

Submarines and ships have copied this method of using sound to find or avoid one another. Ships on the surface can hear the sound of a submarine's engines coming through the water. Fishing boats use sound to find shoals of fish, even when they cannot be seen. This helps them to make good catches.

We use sound travelling through liquid in our own body. The inner ear has a coiled tube full of liquid and when the ear bones vibrate, they push against the end of the tube, making the liquid inside vibrate. The nerves in the tube respond to these vibrations and send impulses to the brain, which 'hears' the sound.

Activity 8: What happens to sounds when they travel through a liquid?

You will need: water; a balloon; a ticking watch, clock or timer; paper (or Workbook) and a pen or pencil.

1 **Work with a partner on this activity.**

   a Ask your partner to hold the watch, clock or timer alongside your head so that you are just able to hear the sound of the ticking.

   b Listen carefully to what it sounds like.

2 **Fill the balloon with water and place it beside your head so that it touches your ear.**

   a Ask your partner to put the watch, clock or timer on the other side of the balloon, so that it is touching it.

   b Listen carefully and compare what you hear this time with the first time.

3 **Swap with your partner and repeat the activity.**

4 **Discuss the observations you both made and come to a conclusion about sound travelling through water.**

The speed of sound varies with the material through which it is travelling. Look at the table below. Sound travels more quickly through solids than through liquids and gases. This is because the atoms in a solid are closer together than in gases and liquids.

Sounds waves travelling at metres per second (m/s) in air, water, concrete and steel

Material (m/s)	Air (at 0°C)	Water	Concrete	Steel
Speed of sound	330	1400	5000	6000

When your ear was off the desk, the sound of the scratching came to you through the air – a gas.

When your ear was on the desk, the sound travelled through the desk – a solid – and it was easier to hear it.

The balloon filled with water made it easier for you to hear the sound of the ticking. It travelled to your ear through the solid wall of the balloon, then the liquid inside it, and finally the solid wall, as the sound moved out and into your ear canal.

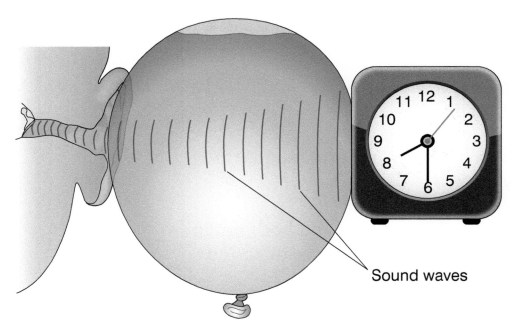

Sound waves

Materials preventing sound transmission

Some materials reduce the **loudness** of sounds, or even completely stop the sound waves from travelling through them.

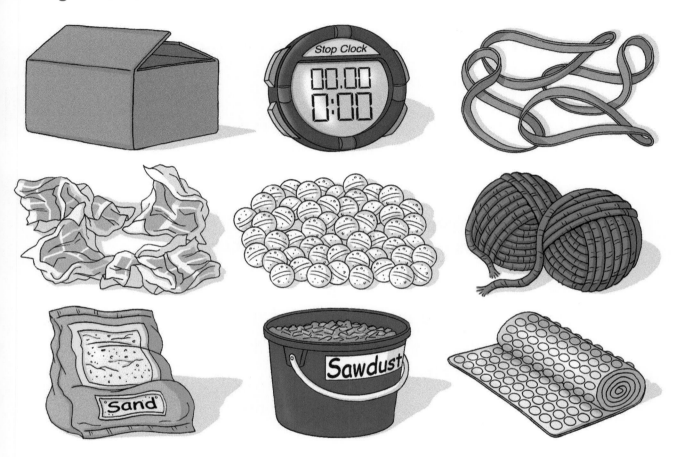

Activity 9: Which material will be best at preventing sound travelling through it?

You will need: a selection of materials to test, a sound source, a sound-level meter, paper (or Workbook) and a pen or pencil.

1 Discuss with your group which materials you think can reduce or prevent sound travelling from its source (its transmission).

Continue over the page

2 **Plan an investigation of how sound transmission can be prevented or reduced.**

a Make it a fair test, using at least *three* materials and a sound source.

b Predict what you think the results will show.

c Write down your prediction.

3 **Decide how you will observe the reduction or prevention of the sound. What will you measure, and how?**

4 **Decide how you will record your observations and prepare for the recording.**

5 **Compare the effects of the three different materials on the transmission of the sound.**

a Carry out the test in a fair way so that you will be able to compare the effects.

b Record your measurements each time.

6 **Compare your results.**

a Use them to make a conclusion about the three materials.

b Compare your results with your prediction.

7 **Share your group's results with the class and compare them with those from other groups.**

Some materials work better than others in preventing sound travelling through them. They can be used as sound **insulators**.

Builders use sound-insulating materials when they want to reduce sounds travelling from one part of a building to another. For example, from one house to another when they share a wall, or from one cinema room to another in a multi-screen cinema.

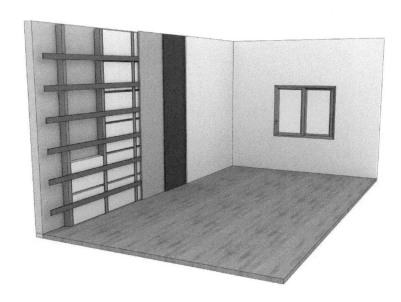

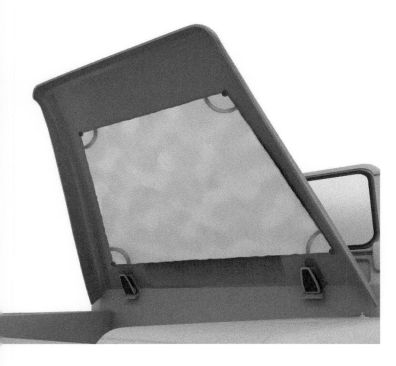

Car makers also use insulating materials under the car bonnet (hood), and on other parts of the car, to reduce the sound of the engine and the noise of the tyres that can be heard inside the car.

At home, sheets of **underlay** material can be put under carpets in upstairs rooms, to reduce the sound heard by people in rooms below.

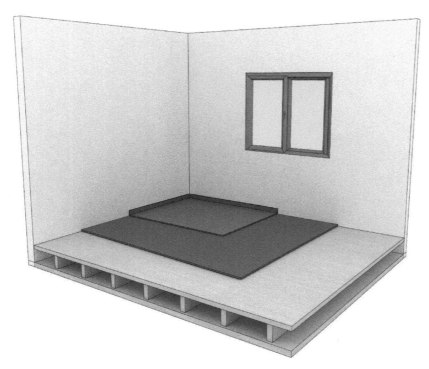

All these sound-insulating materials have many spaces in them between the fibres that form the material. These spaces are filled with air. Sound waves cannot travel directly through these spaces, so they lose energy and the volume of the sound is reduced.

The type of material and the thickness of material both have effects on the amount of sound reduction produced.

Pitch and loudness

The 'highness' and 'lowness' of sounds is called their **pitch**. When we speak, sing, or play certain musical instruments, we can change the pitch of the sounds we make. In this way we can make tunes with our voices and instruments.

Activity 10

You will need: a wooden or plastic ruler, a desk or table, paper (or Workbook) and a pen or pencil.

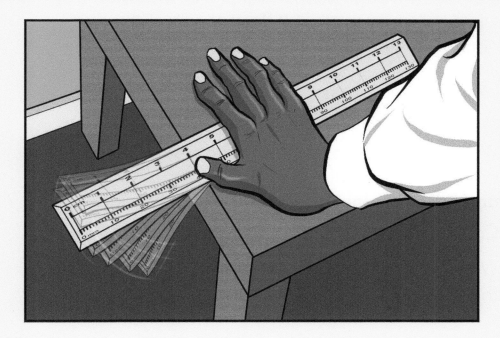

1 Take a ruler or flat strip of wood or plastic.

   a Place it on the edge of the desk so that more than half of it sticks out over the edge.

   b Hold it down firmly.

2 Flick the free end of the ruler. Repeat this several times and listen carefully.

Continue over the page

3 Discuss with your group how you can change the pitch of the sound:

    a to make it a lower sound

    b to make it a higher sound.

 4 Write down in a table what you will do and your prediction of what will happen to the sound.

 5 Test your group's ideas. Record in the table what happens to the pitch of the sound each time you try to change it.

> **Remember** to make it a fair test.

 6 When you have tested all the ideas, compare the results with the predictions and try to explain what you have observed.

    a Can you see a **pattern** in the results?

    b What can you say about the pitch of the sound and its link to the length of the ruler sticking out from the desk?

 7 Share your results and conclusions with the class.

The pitch of sounds can be changed in different ways. What you do with a ruler and a rubber band will not be exactly the same.

Activity 11

You will need: a rubber band, paper (or Workbook) and a pen or pencil.

1 Hold a rubber band between your teeth and stretch it out with one hand, holding it over your thumb and forefinger.

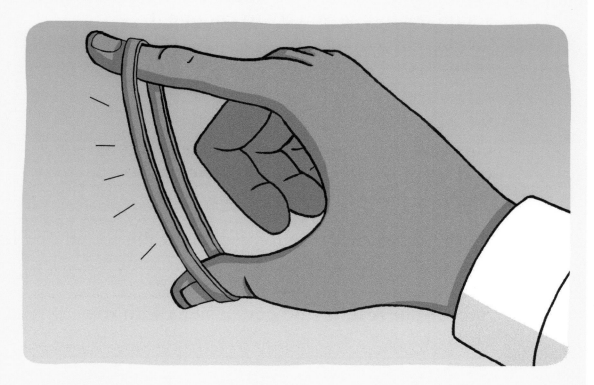

2 Pluck the rubber band with the other forefinger and listen to the sound you have made.

3 Discuss with your group how you can change the pitch of the sound to:

a make it a lower sound

b make it a higher sound.

Continue over the page

4 Write down in a table what you will do, and your prediction of what will happen to the sound.

 5 Test your group's ideas and record in the table what happens to the pitch of the sound each time you try to change it.

> **Remember** to make it a fair test.

 6 When you have tested all the ideas, compare your results with your predictions and try to explain what you have observed.

a Can you see a pattern in your results?

b What can you say about the pitch of the sound and its link to how much you stretched the rubber band?

7 Share your results and conclusions with the class.

Copy and complete the sentences below, using the words listed here:

> higher   longer   lower   shorter

a The pitch of the note gets _____ as the ruler gets _____.

b Making the ruler _____ makes the pitch _____.

c The pitch of the note gets _____ as the rubber band is stretched.

Now add three more sentences of your own about what you discovered when you tried to change the pitch of sounds:

d _____

e _____

f _____

Loudness (volume) is not the same as pitch. A sound can be made louder or softer without changing the pitch.

Sing a note loudly, then softly.

Clap your hands loudly, then softly.

Tap the desk loudly, then softly.

Did the pitch of the sounds change?

Activity 12

You will need: a rubber band, an empty match box or small plastic box, a sound-level meter, paper (or Workbook) and a pen or pencil.

1 Stretch a rubber band over a matchbox or other small container.

2 Pluck the band and listen to the sound it makes.

3 Discuss with your group how you can change the loudness of the sound to make it softer or louder.

4 Write down in a table the group's ideas and predictions of what will happen. You could use these column headings:

Method of making sound	Predicted result	Sound level measurement

5 Test each of the ideas and listen carefully to the sounds you make.

6 Record the results in the table.

7 Compare the results with your predictions and try to explain what you observed.

    a Look for patterns in your results and come to your conclusions.

    b Share your results and conclusions with the class.

Loudness (volume) can be changed by putting more or less energy into the making of the sound.

Sound is a form of energy.

- If you pluck the rubber band harder, the sound is louder.
- If you blow a trumpet harder, the sound is louder.
- If you hit a drum harder, the sound is louder.
- If you shake a shaker more energetically, the sound is louder.

More energy put in produces more energy sent out in the form of sound waves.

A second way to increase the volume is to **amplify** it in some way. For example, the box behind the strings of a guitar, violin or bass, increases the amount of air that vibrates when the instrument is played. This makes the volume louder. Drums use the same method to increase volume.

Musical instruments

People have learned how to make sounds in many ways. For thousands of years, people have been making music. Instruments made from natural materials, such as wood and animal skins, are part of almost every culture across the world.

Strings, metal, seeds and stones are also used in musical instruments to produce sounds.

(a)

(b)

(c)

(d)

(e)

(f)

(g)

(h)

(i)

Look at the pictures of the musical instruments above. Discuss how sound is made by each one. Sort them into four groups and share the groups with the class.

Activity 13

You will need: a musical instrument, paper (or Workbook) and a pen or pencil.

1. Choose an instrument and play with it, exploring how you can change the pitch of its sounds.

2. Record the name of the instrument and the methods that you used successfully to change the pitch of its sounds.

3. Choose a different type of instrument and carry out the same exploration.

4. Record its name and methods as before.

5. Take a third type of instrument and repeat the activity.

6. Compare your results.

   a Come to a conclusion about how the pitch of notes from musical instruments can be changed.

   b Can you see any patterns in your observations?

7. Share your results and conclusions with the class.

As you know, there are different ways to produce a sound, but there must always be a vibration.

Musical instruments all have vibrating parts or vibrating air.

Percussion instruments are hit or shaken. For example, drums, maracas or a piano.

String instruments are plucked or have a bow drawn across the strings. For example, a guitar, violin, oud or harp.

Wind (brass and woodwind) instruments are blown.
For example, a trumpet, flute, alpenhorn or trombone.

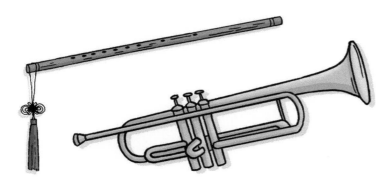

Strings of different length have different pitch.
Strings of different thickness have different pitch.

Wind instruments also make various notes – long tubes and pipes have a different pitch to short ones.

Percussion instruments are sometimes fixed at one pitch. Some, like drums and xylophones, can play notes of different pitches. The length of the wooden keys and the tightness of the skin are related to the pitch of the notes:

• Longer keys and looser skins produce lower notes.
• Shorter keys and tighter skins produce higher notes.

Constructing circuits

Discuss with some other learners what you remember about circuits from Stage 2. Here is a picture of the items needed for making one.

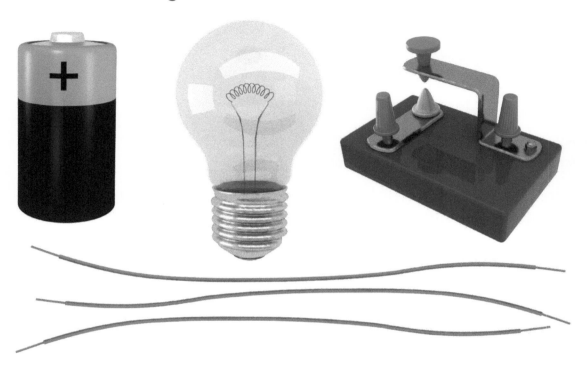

Activity 1

You will need: a battery (cell), a lamp, three insulated wires with bare ends showing, a switch, paper (or Workbook) and a pen or pencil.

1 Collect the items shown in the picture.

2 Try to arrange the six things in a complete **circuit** so that the lamp lights up and can be turned off.

3 Change the way you arrange them.

4 Record in simple drawings all those ways which make the lamp light up and allow you to switch it off.

5 Be careful to show which places on the **battery** and lamp are touching the wires.

6 Try the arrangement shown in the picture:

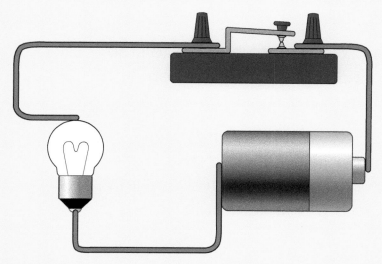

Does it make the lamp light up?

7 Share your results with the class.

The lamp lit only when the switch was closed. If the switch was open, the lamp could not light up.

Broken circuits

When the lamp did *not* light up, this told you that you had *not* made a complete circuit.

There are different reasons why the circuit might not be complete.

- There was a gap, which the electricity could not get across – for example, when the wires were not connected to items correctly.

Or

- The switch was open. We have to be able to stop the **current** flowing and start it flowing when we want it. We use switches to do this.

Or

- The current might have been blocked by a material that did not allow the electricity to travel through it – for example, the glass of the light bulb.

Switches are of many different shapes and sizes. They all do the same thing: they stop and start the flow of current through the circuit.

Activity 2

You will need: electrical items with switches, paper (or Workbook) and a pen or pencil.

1 If you have electrical items at home, look for the switches on them. Make a simple drawing of each kind of switch you find.

2 Continue this activity in school and add more examples to your collection of drawings.

3 Display your drawings for the class to see and look at the drawings of other pupils.

Switches allow us to break and mend circuits by creating a gap.

This gap can be opened (turned off) and closed (turned on). This is a very useful and safe way of controlling electricity.

Switches show that electricity flows only in a complete circuit.

Every time you open a switch, a gap is made and the current stops flowing.

The evidence for this is that, when the switch is turned off, the lamp goes out, the motor stops turning, the bell stops ringing or the buzzer goes silent. Without the flow of current, these items cannot work.

Activity 3

You will need: a battery (cell), a lamp, three insulated wires with bare ends showing, a switch, paper (or Workbook) and a pen or pencil.

1 Rebuild a circuit that lights the lamp.

2 Explore how many ways you can break the circuit and make the lamp go out. Make a simple drawing of each broken circuit.

3 Share your results with the class and explain why each circuit is broken.

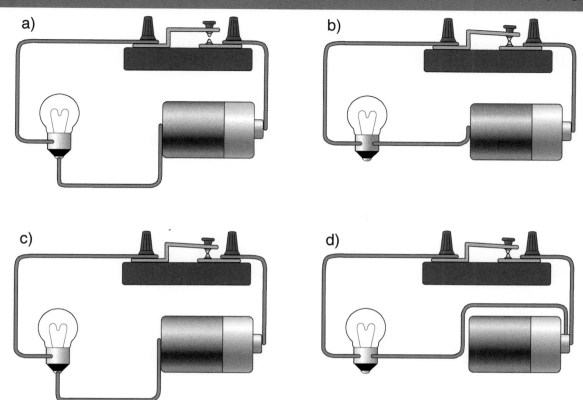

a)   b)

c)   d)

Which of these circuits are complete?

Activity 4

You will need: paper (or Workbook) and a pen or pencil.

1 Look at the drawings of the four circuits.
Which circuits will light the lamp?

2 Discuss your ideas with your group.
Be ready to explain your answers to
the class.

Electrical current

When the current is able to flow, it moves through the circuit.

A circuit is a complete circle that provides a pathway for the electricity to flow out from the battery, through the wires, switch and lamp, and back to the battery.

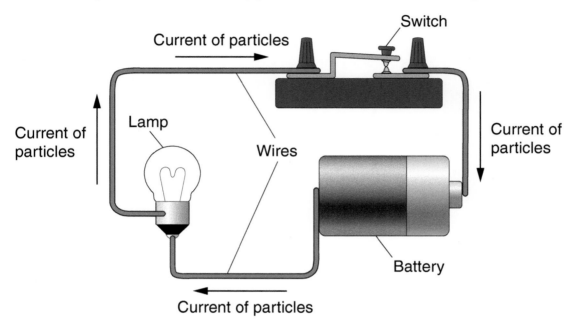

Current of particles

Switch

Current of particles

Lamp

Wires

Current of particles

Current of particles

Battery

If it is complete, the electrical current flows along the wires and through the other items in the circuit. This flow can be thought of as a flow of particles, travelling round and round the pathway of the circuit.

The **filament** of a lamp is heated as the current of particles flows through it. Light is given off from the heated filament; the lamp 'lights up'.

Magnets attract and repel

The bar magnets you have in school are probably painted red at one end and blue at the other. This is done to mark the **poles** of the magnet – the **north** (red) and **south** (blue) poles. Sometimes the letters 'N' and 'S' are also marked on the magnet.

Activity 5

You will need: two bar magnets, some thread, paper (or Workbook) and a pen or pencil.

1 **Tie a thread to one of the bar magnets.**

   a Hold the bar magnet on the thread, letting it swing freely.

   b When it is settled, bring another bar magnet slowly towards it – aiming the north pole of one towards the north pole of the other.

   c Write a note of what you observe.

2 **Repeat the action using the south poles of both magnets.**

   Write a note of what happens this time.

3 **Now bring the south pole of one magnet slowly towards the north pole of the other.**

   Write a note of what you observe.

4 **Lastly, bring the north pole of one magnet towards the south pole of the other.**

   Record what happens this time.

Continue over the page

💬 **5** **Share your results with the class.**

You have found out one property of magnets. This is known as 'the two laws of magnetism':

- Opposite poles **attract**.
- Like poles **repel**.

North and south are opposite poles. A north and a north are like poles. A south and a south are also like poles.

Metals and magnets

Some materials are **magnetic** but what makes them magnetic?

Activity 6: Identifying magnetic materials

You will need: a bar magnet, 10 different objects, paper (or Workbook) and a pen or pencil.

1 Draw a table like this one to record the results of this activity.

Name of object	Material	Magnet attracted	Magnet not attracted

Have spaces for 10 objects.

2 Take a bar magnet and use it to test 10 different objects in the classroom and outside.

a For each object, can you feel the pulling **force** of the magnet (attraction), or not?

b Record the result for each one in the table, with a tick or a cross in the correct column.

Continue over the page

3 **Look at the results.**

a Come to a conclusion about the *materials* attracted by the magnet.

 b Write down your conclusion.

c Share it with the class.

4 **Test your idea about magnetic materials on eight more objects in the room.**

a Write a sentence about the results.

b Compare it with your conclusion from the first test you did.

c Do your conclusions agree, or have you changed your mind about which materials are magnetic?

d Tell the class what you now think.

5 **Sort out these materials into two groups: a magnetic group and a non-magnetic group.**

> glass   plastic   iron   paper   gold   soil
> copper   cloth   rubber   steel   brass   stone
> pottery   bone   silver   tin   wood

Share your groups with the class.

There are many kinds of metals and some of them are found in your classroom and outside.

All the objects that you tested and that you found were attracted by the magnet were _____, but not all _____ were attracted.

What are the missing words?

Only a few metals are magnetic. The rest are non-magnetic.

Iron, steel, cobalt and nickel are the magnetic metals.

Common metals – such as copper, tin, zinc and aluminium – are non-magnetic.

Iron and steel are the most common metals used to make things at home and in school.

They are also used to make tall buildings, vehicles, tools, bridges, railways. These metals are all around us. They are the most magnetic metals.

Howrah Bridge in India

Eiffel Tower in France

Brooklyn Bridge in America

Glossary

A

acid rain – rain that has been made into a weak acid through chemicals contained in smoke.

amplify – to make louder, to increase the volume of a sound.

attract – pull towards.

B

bar chart – a way of showing data in bars or blocks, sometimes on a grid of lines.

battery (electrical) – a device that produces electricity (also known as a 'cell').

boiling – the change of state when a liquid changes into a gas at the boiling point of the liquid.

brass instrument – a musical instrument made of metal, such as brass or silver (for example, a trumpet or trombone).

C

change of state – a physical change from solid, to liquid, to gas or vice versa.

circuit (electrical) – a complete circular route around which electricity flows.

compare – to look for differences and similarities in two or more things or events.

condense – change of state from a gas to a liquid (for example, air to water).

contract – to get smaller or shorter.

current (electrical) – a flow of particles through a circuit.

D

decay – the process of breaking down; rotting the dead bodies of plants and animals.

decibel (dB) – a unit of measurement used to quantify the volume (loudness) of sound.

disposal – getting rid of, removal.

drug – any substance, other than food, that causes a change in the body when swallowed, breathed in, injected or applied to the body (for example, tobacco, alcohol, painkillers or antibiotics).

E

energy – the ability to do work; it is needed to make things happen.

environment – the physical surroundings, including the weather, in which plants and animals live.

evaporate/evaporation – change of state from a liquid to a gas (for example, water to water vapour).

Glossary

evidence – facts, information, proof, clues or data that help us to work something out.

explain/explanation – to give a reason for, tell why it is like it is.

external – on the outside.

F

fair test – a test of an idea in which everything is kept the same except the one thing you are testing.

filament – the fine wire inside a lamp that is heated by the electricity in a circuit, making it glow and give off light.

force – a push or a pull.

freezing – the process of changing from a liquid into a solid (for example, water freezes into ice at 0°C).

G

gas – the state of matter that is not solid or liquid.

H

habitat – the environment that is the natural home of a plant or an animal.

I

insulator/insulation (sound) – a material that prevents or reduces the transmission of sound.

internal – inside.

investigate/investigation – a search for evidence to answer a question.

invisible – cannot be seen (for example, air is invisible).

J

joint – a point in a skeleton where two bones meet and are joined in a way that allows movement (for example, the wrist or elbow).

K

key (biological) – a diagram that shows relationships between different living things, and is used to identify and classify them.

kinetic energy – energy that is the result of movement.

L

ligament – the strong elastic tissue that holds bones together in joints.

liquid – the state of matter that is not solid or gas.

locomotion – movement from one place to another.

loudness – the volume of a sound; a measure of how energetic the sound is.

Glossary

M

magnetic – materials that are attracted by magnets and can be made into magnets (for example, iron, cobalt or steel)

marrow – material inside the long bones in which blood cells are made.

matter – the substances of which the physical universe is made.

measure – to find out the size of an object, a feature or a process (for example, mass, length, time or temperature).

medicines – drugs that help to make us well, or keep us well.

melting – change of state from a solid to a liquid (for example ice to water).

minerals – the materials from which rocks are made. They form most of the soil and are used by plants.

muscles – body tissue that can contract and relax. Bundles of this tissue are attached to bones and produce movement.

N

non-magnetic – materials that are not attracted by magnets (for example, plastic or glass).

O

observe/observation – notice when paying careful attention, for example, when seeing, smelling, hearing, touching or tasting.

P

pattern – some regular feature; for example, a repeated shape, relationship, or measurement.

percussion instrument – an instrument that makes sounds when it is hit or shaken, for example a drum or maraca.

pharmacy/ies – shops where medicines can be bought (prescription and over-the-counter drugs).

pitch – how high or low a note (sound) is.

poles (north/south) – the ends of a magnet where the magnetic force is most powerful.

pollution – damage to the air, soil or water (the environment) with waste materials.

predict/ion – to tell what will happen before doing something.

prescription – a drug used as medicine, which a doctor has chosen as the treatment for an illness.

Glossary

prohibited – against the law; illegal.

protect – to keep safe; to save from being damaged or hurt.

R

recycling – using materials again, rather then throwing them away as waste.

repel – push away; for example, like poles of magnets will push apart when put together.

results – observations of all kinds, including measurements, that are collected during an investigation.

reverse – go back to how it was before something changed.

ribs – bones that form a 'cage' around the chest to protect the lungs and heart.

S

sewage – the solid and liquid human wastes – faeces and urine – that people get rid of in the toilet.

skeleton (internal) – the bones inside the body that support it and form its framework.

skull – in the head, the 'box' of bone that protects the brain inside it.

solid – the state of matter that is not liquid or gas.

sound-level meter – a device used to measure the volume or loudness of a sound in decibels (dB).

spinal column – a flexible set of small bones, running down the middle of the back, from neck to hips, protecting the spinal cord inside it.

states of matter – the three different forms that matter can be in – solid, liquid and gas.

steam – water in the state of an invisible gas at 100°C.

string instrument – a musical instrument played by plucking or drawing a bow across the strings, for example a harp, guitar or violin.

support – hold up; act as a framework.

T

table – a way of writing things down in rows and columns.

temperature – a measure of how hot a substance is.

tendon – the strong tissue that connects muscles to bones.

test – something done to find out if an idea is true or not.

transmission – the process of passing on or sending.

Glossary

U

underlay – sheets of material put on floors under carpets to reduce sounds travelling through them.

V

vibrate/vibration – rapid movement up and down or backwards and forwards; produces sound.

visible – can be seen.

volume – how loud a sound is, measured in decibels (dB), *or* how much space something fills, measured in units such as cubic centimetres (cc).

W

water vapour – water in the state of an invisible gas.

wind instrument – a musical instrument played by blowing air through it.

woodwind instrument – a musical instrument made of wood and played by blowing though it, for example a flute or oboe.